MW01401809

523.48 NAR
Nardo, Don, 1947-
Pluto / by Don Nardo.
04043-612   01-0009695

FLEETWOOD ELEMENTARY SCHOOL LIBRARY
15289 - 88 Avenue
Surrey, B.C. V3S 2S8
Ph: (604) 581-9323, Fax: (604) 581-4481

Eyes on the Sky

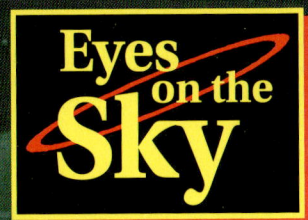

Pluto

by Don Nardo

KIDHAVEN PRESS

THOMSON
GALE

San Diego • Detroit • New York • San Francisco • Cleveland
New Haven, Conn. • Waterville, Maine • London • Munich

THOMSON
GALE

© 2003 by KidHaven Press. KidHaven Press is an imprint of The Gale Group, Inc., a division of Thomson Learning, Inc.

KidHaven™ and Thomson Learning™ are trademarks used herein under license.

For more information, contact
KidHaven Press
27500 Drake Rd.
Farmington Hills, MI 48331-3535
Or you can visit our Internet site at http://www.gale.com

ALL RIGHTS RESERVED.
No part of this work covered by the copyright hereon may be reproduced or used in any form or by any means—graphic, electronic, or mechanical, including photocopying, recording, taping, Web distribution or information storage retrieval systems—without the written permission of the publisher.

LIBRARY OF CONGRESS CATALOGING-IN-PUBLICATION DATA

Nardo, Don, 1947–
  Pluto / by Don Nardo.
  P. cm.—(Eyes on the sky)
  Includes bibliographical references and index.
  Contents: The long search for Planet X—A realm of darkness and ice—Pluto and Charon—the double planet—Is Pluto a major or minor planet?
  ISBN 0-7377-1002-0 (hardback : alk. paper)
  1. Pluto (Planet)—Juvenile literature. [1. Pluto (Planet)]. I. Title. II. Series.
  QB701 .N37 2003
  523.48'2—dc21

2002003594

Printed in the United States of America

Table of Contents

Chapter 1
The Long Search for Planet X 4

Chapter 2
A Realm of Darkness and Ice 14

Chapter 3
Pluto and Charon: The Double Planet . . . 23

Chapter 4
Is Pluto a Major or Minor Planet? 31

Glossary 41
For Further Exploration 43
Index 45
Picture Credits 47
About the Author 48

1
The Long Search for Planet X

Well into the twentieth century, astronomers had confirmed the existence of only eight **planets** in our **solar system**. (The solar system consists of the sun and all the planets, moons, comets, and other objects that move around it.) The planets Mercury, Venus, Mars, Jupiter, and Saturn had been known since ancient times. Uranus and Neptune were found in 1781 and 1846 respectively.

In 1930 a ninth planet—Pluto—joined the sun's cosmic family. **Orbiting** slowly in the dark reaches of the outer solar system, Pluto was and remains the least understood and most mysterious planet. Also, its discovery was a long and difficult process. It took many years and involved much educated guesswork, complex mathematical calculations, patience, and just plain hard work.

Planets Beyond Uranus?

The long search for Pluto began shortly after the 1781 discovery of Uranus by English astronomer William Herschel. At first he and his fellow astronomers assumed that Uranus marked the outer edge of the solar system. But eventually some of these men began to suspect that at least one more planet lay beyond it.

This idea was based on Uranus's odd behavior. Researchers had watched it move and calculated its orbit, or path around the sun. But over time the planet strayed farther and farther from this predicted orbit. By 1834 a number of astronomers were convinced that an unknown planet existed beyond Uranus. The **gravity** of this

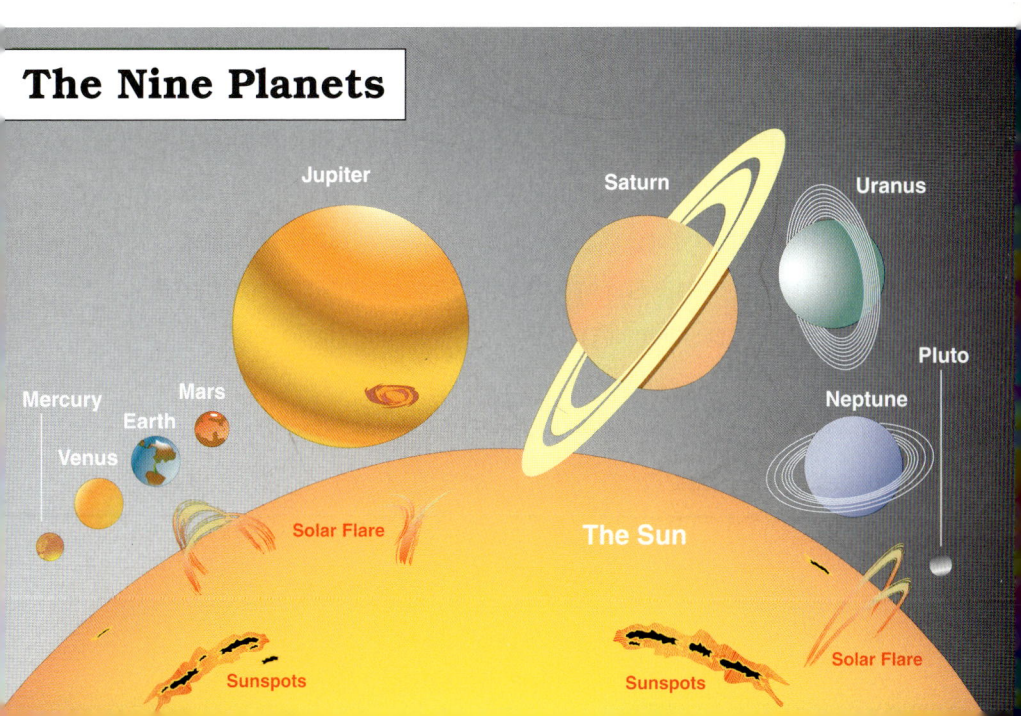

The Nine Planets

planet was pulling on Uranus, they proposed, keeping it from following its predicted orbit.

In the years that followed, several researchers tried to locate the new planet using mathematics. They carefully observed Uranus's movements. Then they measured how much these movements differed from those that had been predicted. Finally, they calculated the location of the unknown planet based on the amount of this difference. Two men—Englishman John Couch Adams and Frenchman Urbain J.J. Leverrier—attacked this problem independently. And both worked out roughly the same position in the sky for the new planet. Sure enough, when an astronomer pointed his telescope at that position in September 1846 he saw a small, fuzzy, bluish-colored disk. Astronomers named the new planet Neptune, after the Roman god of the seas.

Neptune's discovery was one of the great achievements of modern science. But the question of which planet marked the solar system's outer boundary remained. As the years went by, some researchers noted that Neptune's gravity accounted for only part of Uranus's odd behavior. The suspicion began to grow that still another unknown planet existed.

Planet O and Planet X

One researcher who was convinced that a planet lay beyond Neptune was David Peck

The eighth planet, Neptune, as photographed by a modern ground-based telescope.

Todd of Amherst College in Massachusetts. He did a series of calculations similar to those of Adams and Leverrier, and in 1877 Todd thought he had the answer. The unknown planet orbited the sun at a distance of fifty-two **astronomical units**, he proposed. (An astronomical unit, abbreviated AU, is the distance from Earth to the sun, or about 93 million miles. For convenience, astronomers use AUs to express large distances within the solar system.) Also, said Todd, the planet took 395

An artist's conception of Planet O (upper left), with two imaginary moons nearby.

Earth years to go around the sun once. However, Todd was greatly disappointed when he was unable to locate the planet in the area predicted.

Two well-known American astronomers soon picked up the search where Todd had left off. The first was William Henry Pickering, who called the object of the search "Planet O." By 1908 he had worked out an orbit similar to Todd's—a distance of 51.9 AUs and a period of 373.5 years. But Pickering suggested a different position in the sky than Todd had. Unfortunately for Pickering, a search of the area he had indicated proved fruitless.

Meanwhile, Percival Lowell of Lowell Observatory in Flagstaff, Arizona, was working on the problem at the same time as Pickering. In recent years Lowell had been studying the planet Mars. He claimed he saw canals there, and he was convinced that intelligent beings had built these waterways. Many other astronomers thought these ideas had no merit and ridiculed Lowell. He joined the quest for the ninth planet partly to prove himself to his critics.

In fact, Lowell became obsessed with finding the unknown planet, which he called "Planet X." But like Pickering, he failed to locate it. Lowell died in 1916 unaware of how close he had come. The year before, one of his assistants, Thomas Gill, had actually photographed the planet. But it had appeared as only one of thousands of tiny dots on the photographic plate, and the men who examined the plate had missed it.

The Would-be Astronomer

In the twelve years following Lowell's death his successors at Flagstaff were beset by legal and financial problems, so they put the search for Planet X on hold. By 1928, however, the three astronomers then on the observatory's staff decided they wanted to resume the search. But there was a problem. Examining one photographic plate after another and trying to pick out one pinpoint of light from a multitude of others was a very time-consuming task. And they were all too busy with other projects to tackle the job.

Then a very timely letter arrived at the observatory. It was from a young Kansas farmer named Clyde W. Tombaugh, who wanted to become an astronomer. He explained how he had taught himself about the stars and planets. He had also built his own telescope out of spare auto parts and other makeshift materials. Impressed, the director of the observatory decided to hire Tombaugh as an assistant. The eager young man seemed like a perfect candidate to do the laborious job of searching for Planet X.

Slow and Tiring Work

Tombaugh, then twenty-four, arrived at Flagstaff in January 1929 and began the search in April. His main tool was a device called a **blink comparator**. Most earlier researchers had examined photographic plates with a magnifying glass, a very slow and tiring process. Not surprisingly, after looking at images of thousands of stars for several hours, one could easily make mistakes.

Working with the blink comparator was faster and more precise. The operator placed two photos side by side in the device. Each photo covered the exact same portion of the

Clyde Tombaugh gazes into a blink comparator, the device he used to find Pluto.

sky, except that one was taken a few days or more after the other. In that short time span the stars appeared to be motionless, so their images remained in the same positions in both photos. By contrast, the position of a planet or other moving object changed from one photo to the other. The device blinked a light on and off, alternating between the photos in a fraction of a second. Any image that had moved appeared to jump up and down or from side to side.

Though considerably better than a magnifying glass, using the blink comparator was still slow and tiring work. Tombaugh later called it the most boring job he had ever done. But he stuck with it diligently day after day, month after month. From time to time an image would jump when the light blinked, revealing an object moving through the sky. But it would turn out to be a comet or an **asteroid**. The young man was not unhappy about finding these mountain-sized chunks of rock and ice; but he was out for bigger game, namely Planet X.

Success at Last

Finally, Tombaugh's dogged patience and hard work paid off. On February 18, 1930, he was blinking some photographic plates taken the preceding month. Suddenly, one of the star images jumped. Was it merely another asteroid?

An artist's view shows Pluto's surface as heavily cratered. No one knows yet if this is accurate.

Closer examination of the plates, along with some brief calculations, showed that it was not. First, the object moved more like a planet than an asteroid; also, it was located well beyond Neptune, at roughly the distance astronomers expected a ninth planet to be.

There was no doubt in Tombaugh's mind that he had found Planet X. After several days of checking and rechecking, the older Flagstaff astronomers agreed. On March 13 they announced the discovery to the world. The press and public alike seemed to think the name the scientists had chosen for the new planet was very fitting. Pluto was the Roman god of the Underworld, a dark and gloomy realm. Clearly the new planet dwelled in a place no less dim and foreboding.

2

A Realm of Darkness and Ice

Perhaps someday astronauts from Earth will visit Pluto. If so, the first thing they will notice is how dimly lit the planet is. It lies at an average distance of 39.5 AUs from the sun, nearly forty times farther out than Earth. To a person standing on Pluto's surface, therefore, the sun would seem very remote indeed. It would not appear as a blindingly bright disk, as it does from Earth; instead, the sun would look no more impressive than an unusually bright star in the inky black sky.

Because Pluto orbits so far from the sun, it also receives extremely little warmth; so not surprisingly, the planet is very, very cold. Even the daytime temperatures can plummet lower than -380 degrees Fahrenheit. That is more than 410 degrees lower than the temperature at which water freezes into ice. Therefore, it is more than fitting to say that Pluto rules a realm of darkness and ice.

Too Far for Close-up Views

While the sun looks small and distant from Pluto, to the unaided eye Earth is not visible at all. Similarly, it is not possible to see Pluto with the naked eye from Earth. Even in the largest ground-based telescopes, the outermost planet appears as a tiny white blob. This is partly because Earth's atmosphere distorts light, including images of the **celestial**, or heavenly, bodies.

Pluto, shown at lower left, is so far from the sun that the planet receives little sunlight.

In contrast the Hubble Space Telescope (HST) orbits above the atmosphere, so it produces much sharper images than ground-based instruments. In photos taken by the HST Pluto is a distinct disk. Yet no surface details are visible. The planet is just too far away to allow humans close-up views like those the HST has taken of the moon, Mars, and Jupiter.

Because Pluto is so far from the sun, the planet's orbit is extremely large. Pluto travels along its orbital path at three miles (five kilometers) per second, which by human standards is very fast. Yet it still takes about 248 Earth years to go around the sun once. This means that a single year on Pluto lasts 248

The Hubble Space Telescope orbits above Earth's atmosphere, snapping extremely clear photos.

Earth years. It also means that Pluto will not return to the place where Clyde Tombaugh discovered it in 1930 until the year 2178!

The Alternating Title of Outermost Planet

Pluto's orbit is not only distant and large, but also quite distorted as compared to the orbits of the other planets. First, Pluto's orbit is more tilted than the others. Most of the planets travel around the sun in the same plane, much as Saturn's rings all lie in a flat disk around that planet. However, Pluto's orbit is tilted at an angle of seventeen degrees from the others.

Also, Pluto's orbit is somewhat off-center as compared to the other planetary orbits. At times Pluto lies far beyond Neptune at a distance of about fifty AUs from the sun. But over time the smaller planet moves as close as thirty AUs. In fact, for a period of twenty years in each orbit Pluto moves close enough to the sun to lie slightly inside Neptune's orbit.

This creates a strange situation in which Pluto temporarily loses its title as the outermost planet. Such was the case from 1979 to 1999. During those years Neptune was the outermost planet in the solar system. Pluto has since regained that title and is presently drifting back into the dark gulfs beyond Neptune.

A Realm of Darkness and Ice

Pluto's Orbit

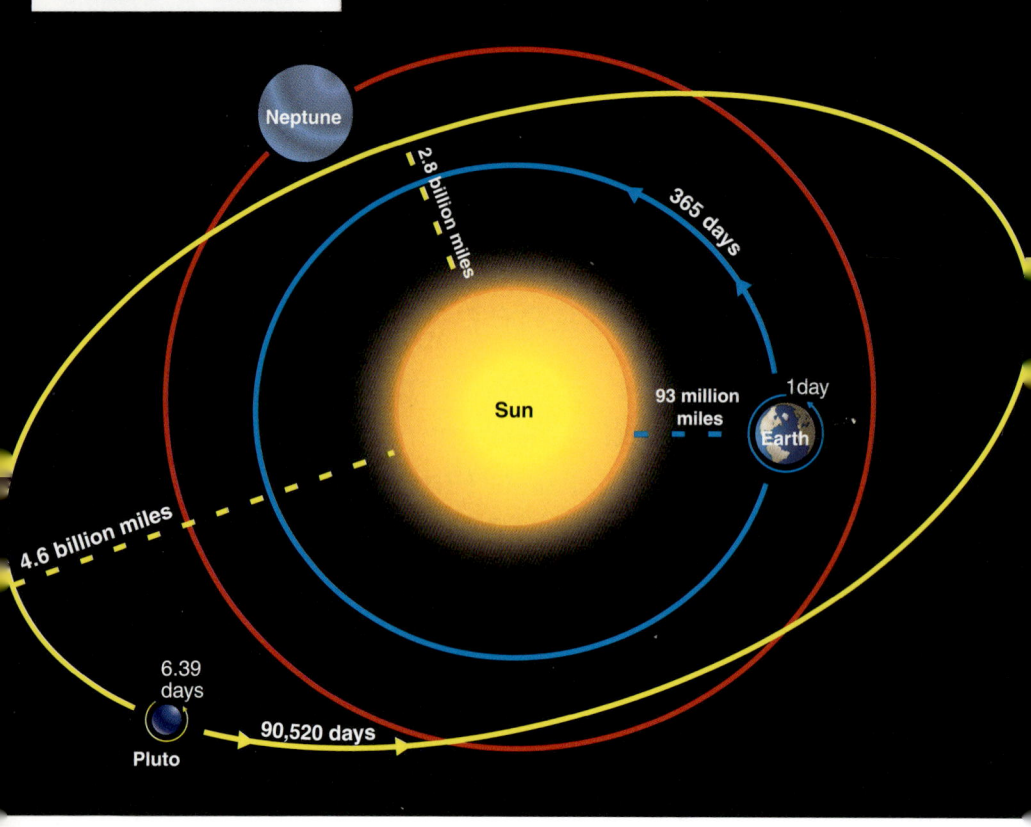

Since Pluto sometimes crosses Neptune's orbit, it is natural to wonder whether the two planets might eventually collide. Luckily, the likelihood of such an event is extremely low. Because Pluto's orbit is tilted considerably from Neptune's, the two never come closer than several million miles.

Size and Composition

Another reason that photos of Pluto taken from Earth reveal little detail is that Pluto is very small. In fact, Pluto is by far the smallest planet in the solar system. Astronomers estimate that

18   Pluto

its diameter is only about one-fifth that of Earth, or two-thirds that of the moon. The next biggest planet, Mercury, is twenty-five times larger than Pluto. And Pluto is only about ten times bigger than the largest asteroid, Ceres.

Pluto's **mass** is also very small. It would take about five hundred objects the size of Pluto to equal Earth's weight. This means that the materials Pluto is made of are fairly light. It does not have a heavy iron core as Earth does, for example. Instead, Pluto has a rocky core.

Surrounding this rocky core are layers of ices. Besides frozen water these include frozen ammonia and methane, poisonous substances that are in gaseous form on Earth. In addition, evidence shows that like Earth and Mars, Pluto also has northern and southern polar caps. The difference is that Pluto's are composed of frozen nitrogen, an element abundant in Earth's atmosphere; small amounts of frozen methane also exist in Pluto's polar caps.

Pluto's Atmosphere

Pluto's polar caps seem to have formed by falling nitrogen and methane snow. This reflects the fact that the planet has an atmosphere made up of these substances. The first direct evidence for this atmosphere came in June 1988 when Pluto passed in front of a bright star.

Pluto's Composition

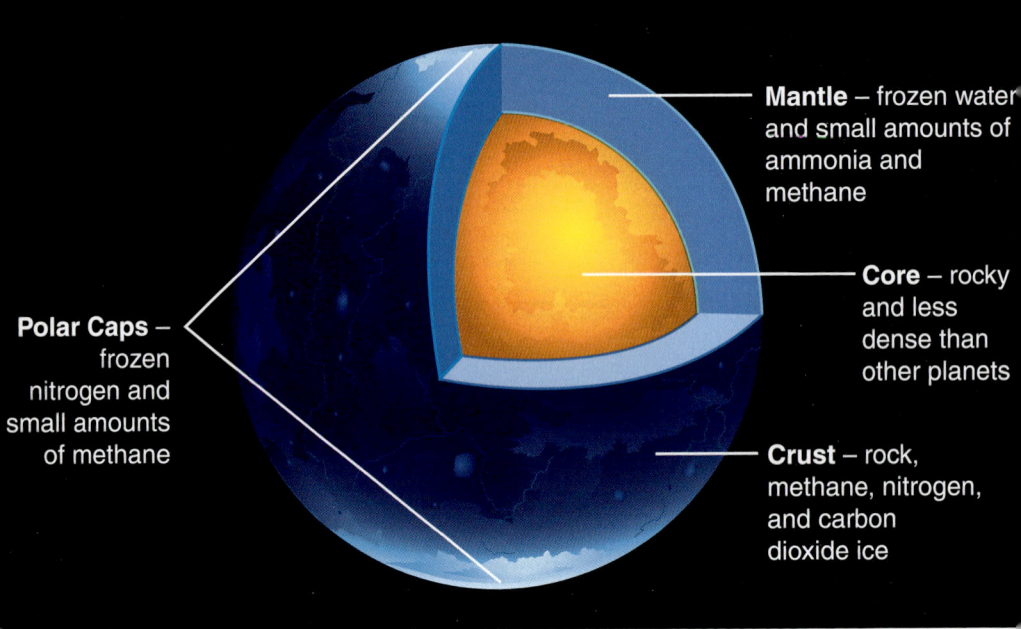

Mantle – frozen water and small amounts of ammonia and methane

Core – rocky and less dense than other planets

Polar Caps – frozen nitrogen and small amounts of methane

Crust – rock, methane, nitrogen, and carbon dioxide ice

For a brief moment just before the star disappeared behind the planet, the star's light lit up the Plutonian atmosphere. Using instruments on Earth, astronomers were able to measure the atmosphere's temperature and chemical makeup.

These measurements also revealed that Pluto's air is extremely thin. Experts estimate that the planet's atmosphere is at best only one-hundred-thousandth as dense as Earth's atmosphere. So if a person stood on Pluto without a spacesuit, the thinness of the air (not to mention the lack of oxygen and freezing temperatures) would kill him or her in seconds.

An ongoing relationship exists between Pluto's atmosphere and polar caps. When Pluto comes closest to the sun, as it did in 1989, the planet receives a little more sunlight than usual. The extra warmth causes some of the frozen methane in the polar caps to change into a gas. So for a while, the atmosphere becomes more substantial. But as Pluto begins moving farther from the sun, the methane starts freezing again; and in the form of snow it

Pluto's polar caps are visible in these images taken by the Hubble Space Telescope.

Pluto
PRC96-09a · ST ScI OPO · March 7, 1996 · A. Stern (SwRI), M. Buie (Lowell), NASA, ESA
HST · FOC

falls back onto the surface, replenishing the polar caps.

This natural cycle has gone on for billions of years. Barring some unexpected disaster, it will continue uninterrupted for billions more. Endless regularity and sameness are the rule in that faraway silent realm of darkness and ice.

3
Pluto and Charon: The Double Planet

Although science learned of the existence of Pluto in 1930, almost half a century elapsed before the discovery of its single moon—Charon (pronounced KAIR-on). It is not surprising that it took so long to find Charon. The moon and its parent planet lie unusually close together. Because they are extremely distant from Earth, for a long time they appeared as a single object in Earth-based telescopes.

Charon is also unusually large in relation to its parent. Earth's own satellite, the moon, has a similar distinction. For that reason the Earth-moon system was and still is sometimes referred to as a double planet. However, the Pluto-Charon system is a good deal more deserving of the title. Charon is by far the largest

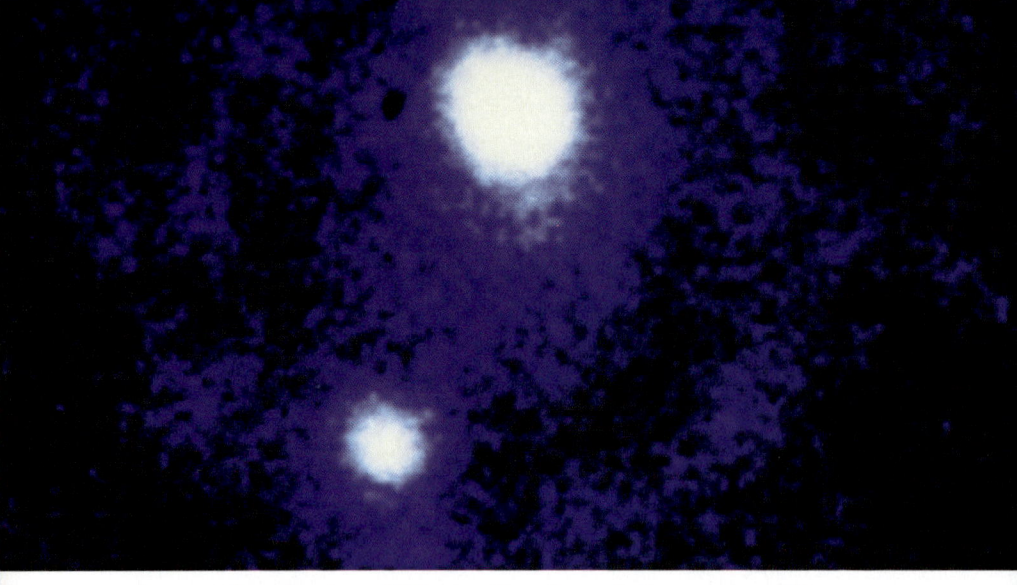

An early photo of Pluto (top) and its moon Charon shows them as fuzzy blobs.

satellite in the solar system in comparison to its parent planet.

Charon's Discovery

That a double planet exists at the edge of the solar system came as a big surprise to astronomers. Pluto is very small as planets go, so they assumed that any moons it might have would be extremely small, too. In any case, they reasoned, a large satellite would already have shown up in photos. Almost no one suspected the truth—that a large moon lay very close to Pluto. Even in the largest telescopes the disks of the two objects had long merged into one fuzzy blob.

One day in 1978 researcher James W. Christy of the U.S. Naval Observatory was photographing that fuzzy blob. His goal was to increase knowledge about Pluto's orbit. But when he examined the photographic plates, he noticed something odd. The planet's ill-defined disk appeared to bulge slightly on one side. At first Christy thought that someone had bumped the telescope while the camera was working. That could cause the planet's image on the plates to move or blur.

But then Christy saw that the images of the stars surrounding Pluto's disk were round and sharp. Any jarring of the instrument would have distorted their shapes, too. He photographed the planet again. And once more he saw the bulge. Only this time it had moved along the edge of Pluto's disk. More pictures taken over the course of a few weeks showed that the bulge circled the planet's image once every six days and nine hours.

This evidence led Christy to believe that Pluto had a satellite that lay too close to the planet to see as a separate body. Other evidence

A more recent photo of Pluto (left) and Charon shows them more clearly as spheres.

later confirmed his suspicion. It became clear that Pluto does indeed have a moon, and that moon orbits the planet once every six days and nine hours. (Since that time, the Hubble Space Telescope and some new, very large ground-based telescopes have photographed Pluto and Charon. These instruments are powerful enough to show them as two distinct disks.)

Once the newly found satellite's existence was verified, the next step was to name it. Astronomers thought it would be appropriate to choose a name related to the mythical Pluto and his underground realm, so they named the moon after Pluto's grim boatman. In Greco-Roman legend, Charon ferried people's souls across the river that marked the boundary of the Underworld.

A nineteenth-century painting depicts Charon, the mythical boatman of the Underworld.

Charon's Diameter and Mass

Over time, researchers compiled other data about Pluto and Charon and their unique relationship. First, their sizes relative to each other seemed to justify calling them a double planet. Careful observation and measurement showed that Charon is about 750 miles (1,200 kilometers) in diameter. That is 52 percent the diameter of Pluto. Also, Charon's mass is almost a quarter of Pluto's mass. By comparison, the moon's diameter is less than 30 percent that of Earth, and the moon is less than an eightieth as massive as Earth. These figures confirm that Pluto-Charon is much more of a double planet than Earth-moon.

Astronomers also measured the distance between Pluto and Charon. They found it to be about 12,000 miles (20,000 kilometers). That is only 5 percent, or one-twentieth, the distance from Earth to the moon. Put another way, it is only one-and-a-half times Earth's diameter. It is no wonder, then, that Pluto and Charon long blurred together into a single object in Earth's telescopes.

Locked in a Weird Embrace

This unusual closeness between Pluto and its moon is also responsible for a strange phenomenon in the Plutonian system. Over time, the

Pluto and Charon Versus Earth and Moon

Pluto — 1,429 miles — 1/50 mass of Earth

Side of Pluto always faces Charon

Side of Charon always faces Pluto

Charon — 758 miles — 1/2,000 mass of Earth

6 days 9 hours to revolve around Pluto

12,000 miles

Earth — 7,909 miles

Side of the moon always faces Earth

Moon — 2,140 miles — 1/81 mass of Earth

1 day to revolve around Earth

238,900 miles

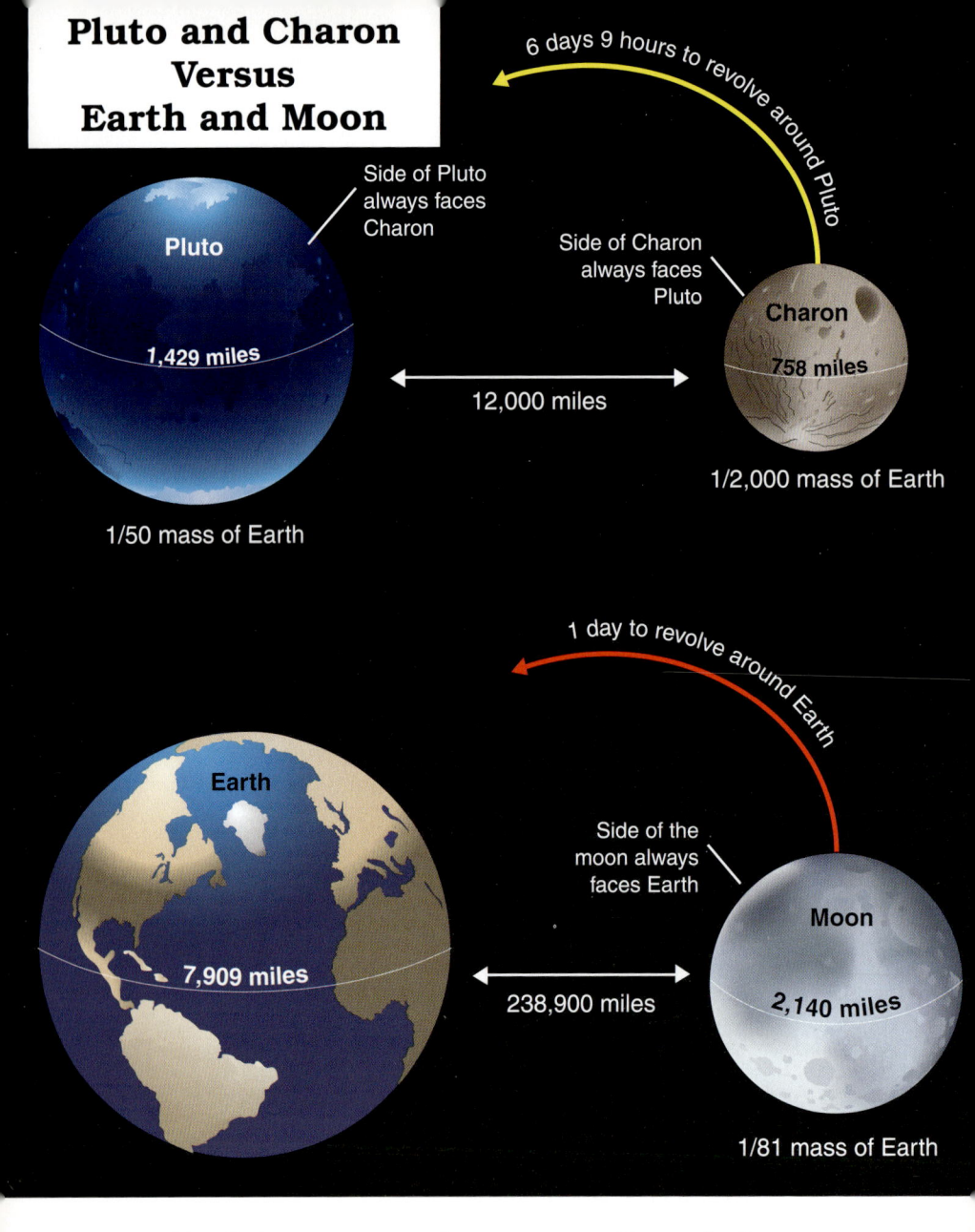

pull of their mutual gravities slowed the rate at which each **rotated**, or spun on its axis. Finally, Charon stopped rotating entirely and now eternally keeps the same side facing Pluto. (The same thing happened to Earth's satellite; hu-

28   Pluto

mans never saw the far side of the moon until spacecraft flew around and photographed it.)

Meanwhile, Charon's movement around Pluto has become locked in time with the planet's own rotation period, or its day. The moon orbits its parent in six days and nine hours, and Pluto rotates once in the same amount of time. That means that a person living on one side of Pluto would always see Charon hanging motionless in the sky; while someone inhabiting the other side of the planet would never see Charon at all.

Astronomers say the same thing will happen someday to Earth and moon. Earth's rotation is

The sun lights up the edges of Pluto and Charon in the solar system's dim outer reaches.

slowing. And Earth's days are growing longer by a very tiny amount each century. Many millions of years from now, Earth and the moon will be locked in the same weird embrace as Pluto and Charon. A day on Earth will last fifty times longer than it does today, and the moon will move around Earth in the same period—fifty days. Thus, studying the strange relationship between Pluto and Charon provides humanity with an unsettling preview of its own future.

4

Is Pluto a Major or Minor Planet?

Pluto has been called the ninth planet in the solar system ever since its discovery in 1930. But in recent years a number of astronomers and other researchers have questioned whether it warrants the status of planet. At least, they say, Pluto should not be classified as a *major* planet. According to this view, Pluto is too different from major planets like Jupiter, Neptune, Earth, and Venus. Other researchers disagree. They contend that Pluto's planetary status is well deserved and should remain intact.

Too Small for a Planet?

Lying at the heart of this argument is the question of just what a planet is. The generally

This diagram shows how tiny Pluto is when compared to the other planets.

accepted definition is a large, **spherical** (round) object that orbits a star independently and shines by reflected, rather than its own, light. Those who accept Pluto's present planetary status say it fits this definition perfectly well.

But their opponents say that Pluto fulfills only some of the standards of the definition. They point out that Pluto has certain physical properties that disqualify it as a major planet. First, they argue, Pluto is too small in comparison to the other planets. Until Pluto's discovery, Mercury held the title of smallest planet in the sun's family. And it turns out that Pluto is twenty-five times smaller than Mercury.

A similar argument is that Pluto does not qualify as a planet because it is smaller than seven of the moons in the solar system. These include Earth's moon, Jupiter's four largest satellites, Saturn's biggest moon, and Neptune's largest moon, Triton. Therefore, this argument goes, Pluto's size is more like that of an asteroid than a planet. An asteroid is a hunk of rock and/or metal that orbits the sun. Asteroids range in size from a few hundred feet to several hundred miles across.

Those who disagree with this view say that the definition for a planet does not specify any exact size. It merely says "large." But how large is large? They admit that Pluto is a good deal smaller than Mercury, but they point out that Pluto is far larger than the biggest known asteroid, Ceres; so it would be mistaken to call Pluto an asteroid. Most important, Pluto orbits the sun independently, is spherical, and shines by reflected light. Therefore it is a full-fledged planet.

Mistaken for a Mystery Planet

Another argument against Pluto's status as a major planet involves its discovery process. This view recalls that Clyde Tombaugh found it while searching for Planet X; yet Pluto is *not* that elusive object. Astronomers had proposed the existence of Planet X to explain small variations in the orbits of Uranus and Neptune. But Pluto's

An artist's conception of Ceres, the largest known asteroid. Ceres is a good deal smaller than Pluto.

gravitational pull turned out to be too small to cause these variations. Furthermore, most astronomers now believe these variations were caused by human error. That means there may never have been a Planet X at all. And if so, Pluto's discovery was merely a matter of fortunate chance.

Those who accept this view ask what would happen if Pluto was discovered today, when no one is searching for Planet X. In that case, they say, Pluto would not be mistaken for a mystery planet. Therefore it would not be classified as a major planet. Instead, because of its tiny size, it would be called a minor planet, as are the larger asteroids. In 1998, in fact, a number of leading international astronomers proposed that Pluto's status be changed to that of minor planet. But others, including most American astronomers, strongly objected. So no action was taken.

Was Pluto Once a Moon?

These and other arguments over whether Pluto qualifies as a major planet are directly related to the mystery of Pluto's origins. Indeed, astronomers have long wondered where Pluto came from in the first place. Was it originally a moon, an asteroid, a comet, or some other body? The answer to these questions may well affect the way science views and classifies Pluto in the future.

Astronomers have never agreed completely about Pluto's origins. But for many years the most popular theory was that it began as a satellite of Neptune. In fact, Pluto does bear a strong physical resemblance to Neptune's moon Triton. According to this view, some sort of cosmic disaster ripped Pluto away from its parent and hurled it into an independent orbit around the sun.

Pluto closely resembles Triton, seen in this photo taken by the *Voyager 2* spacecraft in 1989.

Perhaps Triton itself was the culprit. Many astronomers suspect that it formed as an independent planet; that the gravity of some larger body pushed it out of orbit, and Neptune's gravity captured it. If so, Triton's entry into the Neptunian system could have dislodged Pluto and sent it packing. In this scenario Pluto began as a moon and ended up as a planet, while Triton began as a planet and ended up as a moon.

A New Class of Cosmic Bodies

More recently many astronomers have come to believe that Pluto formed in the dark void beyond Neptune. In 1992 the first of a new class of cosmic bodies was discovered. They are called **Kuiper Belt Objects (KBOs)**, after Gerard Kuiper, the astronomer who predicted them in the 1950s. Pluto strongly resembles these objects and may be the largest of their number.

The **Kuiper Belt** is essentially a second **asteroid belt**, similar to the one that lies between the orbits of Jupiter and Mars—except that the KBOs lie beyond Neptune, between about thirty and fifty-five AUs from the sun. About five hundred of these asteroids have been found so far;

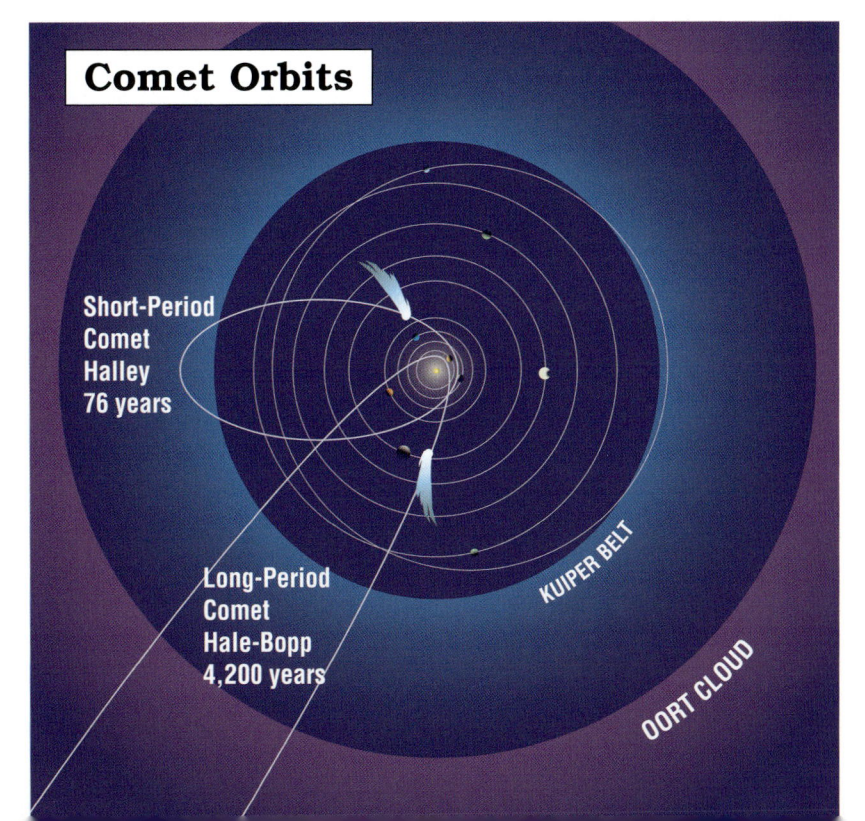

Comet Orbits

Short-Period Comet Halley 76 years

Long-Period Comet Hale-Bopp 4,200 years

KUIPER BELT

OORT CLOUD

and astronomers believe that millions or even billions of others exist.

The consensus of experts is that the KBOs are **planetesimals**. These are the remainders of chunks of rock, metal, and ice that combined to form the planets in the early solar system. After the completion of planetary formation, some planetesimals were left over and became the asteroids and comets that still orbit the sun. If Pluto is indeed a KBO, it began as a planetesimal and became a large asteroid. That would strengthen the argument for reclassifying it as a minor planet.

Debris floats through space after the explosion of a star. Charon may have formed from such debris.

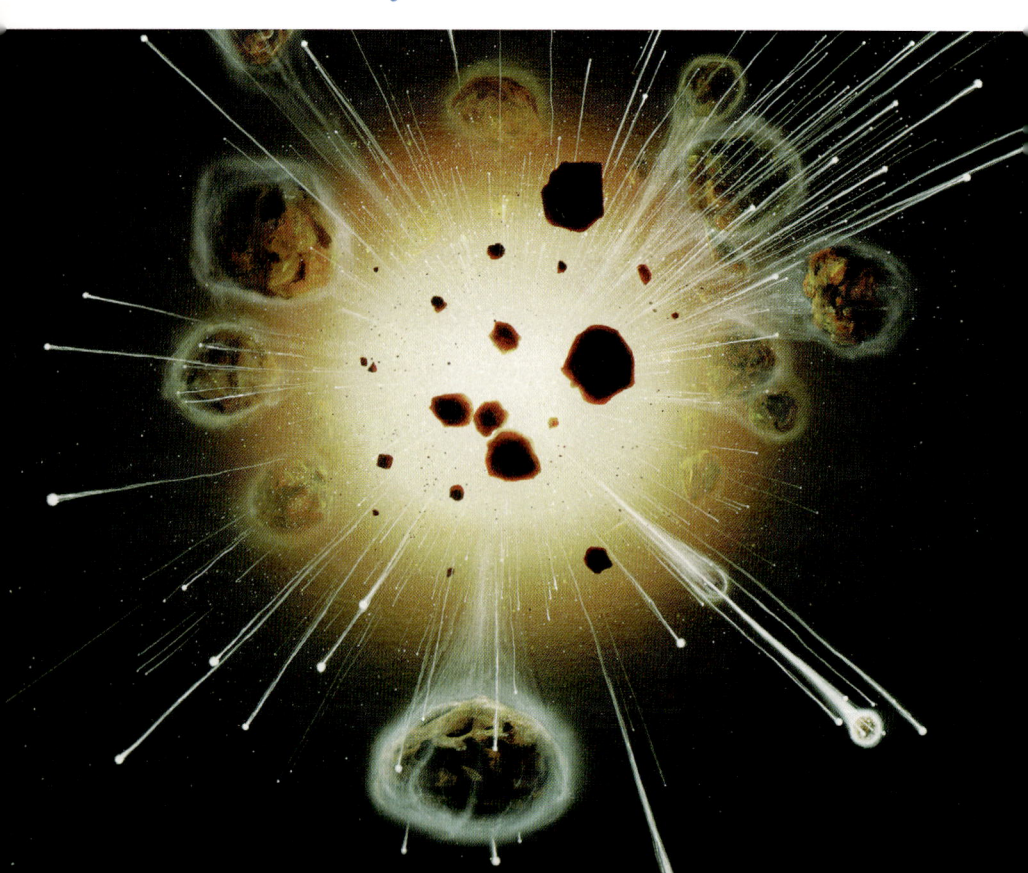

Earth-Centered Universe

As for Charon, it too could have started out as a KBO. Eventually, it may have ventured too close to Pluto and the gravity of the larger object captured it. However, most astronomers favor a different origin for Charon. They postulate that it formed when a large object, probably another KBO, crashed into Pluto. The impact blasted large amounts of debris into space. Much of this material remained in orbit around Pluto and slowly came together to form a moon—Charon. Astronomers believe Earth's moon formed in a similar manner after a Mars-sized object collided with Earth.

The Nature of Science

It remains unclear whether the question of Pluto and Charon's origins or other factors will affect Pluto's planetary status. Pluto may well be reclassified as a minor planet in the future. If so, it should be seen less as a demotion and more as a normal part of the scientific process. The very nature of science demands that ideas must change in the face of new evidence.

Changing Ideas

Indeed, the history of astronomy is full of examples of ideas and views changing and evolving. People used to think the sun moved around Earth until evidence showed the reverse to be true. And astronomers long thought the craters on the moon were caused by volcanoes; now they know that most formed from the impacts of asteroids and other space debris.

Similarly, ideas about what Pluto is and how it should be classified are beginning to change. Yet such changes can never diminish the importance of Pluto and its unique moon as cosmic bodies. Surely the Pluto-Charon system will always remain one of the most fascinating and frequently studied entities in the solar system.

Glossary

asteroid: A small rocky or metallic body orbiting the sun, most often in the asteroid belt or Kuiper Belt.

asteroid belt: A region lying between the orbits of the planets Mars and Jupiter, where many of the asteroids are located.

astronomical unit (AU): The distance from Earth to the sun, or about 93 million miles.

blink comparator: A device long used by astronomers to find the image of a planet or other moving object on a photographic plate containing the images of thousands of stars.

celestial: Heavenly; located in the sky.

gravity: A force exerted by an object that attracts other objects. The pull of Earth's gravity keeps rocks, people, and houses from floating away into space. It also holds the moon in its orbit around Earth.

Kuiper Belt: A region lying beyond Neptune and containing millions or billions of asteroids or similar bodies.

Kuiper Belt Object (KBO): An asteroid or similar body orbiting in the Kuiper Belt. A number of astronomers think Pluto may be a large KBO.

mass: The total amount of matter contained in an object.

orbit: To move around something; or the path taken by a planet or other heavenly body around the sun, or a moon around a planet.

planet: A large spherical object that orbits a star independently and shines by reflected, rather than its own, light.

planetesimals: Small objects that orbited the early sun and combined to form the planets.

rotate: To spin around a central axis (imaginary pole in the center).

solar system: The sun and all the planets, moons, asteroids, and other objects held by the sun's gravity.

spherical: Round like a ball.

For Further Exploration

Pam Beasant, *1000 Facts About Space.* New York: Kingfisher Books, 1992. An informative collection of basic facts about the stars, planets, asteroids, and other heavenly bodies.

Carmen Bredeson, *Pluto.* Danbury, CT: Franklin Watts, 2001. Explains the basic facts about Pluto and its moon, Charon, to young readers.

Kenneth C. Davis, *I Don't Know Much About the Solar System.* New York: HarperCollins, 2001. An exciting fact-filled tour of the solar system, including Pluto, set in a question-and-answer format.

Nigel Henbest, *DK Space Encyclopedia.* London: Dorling Kindersley, 1999. This beautifully mounted and critically acclaimed book is the best general source available for grade school readers about the wonders of space.

Robin Kerrod, *The Children's Space Atlas: A Voyage of Discovery for Young Astronauts.* Brookfield, CT: Millbrook Press, 1992. A well-written informative explanation of the stars, planets, comets, asteroids, and other objects making up the universe.

Don Nardo, *Neptune.* San Diego: KidHaven Press, 2002. This colorfully illustrated book tells the dramatic story of how Neptune was discovered, which had a major bearing on the search for Pluto.

———, *The Solar System.* San Diego: Kid-Haven Press, 2003. An informative overview of the sun's family, including the outermost planet, Pluto.

Gregory L. Vogt, *Asteroids, Comets, and Meteors.* Brookfield, CT: Millbrook Press, 1996. Tells the basic facts about these stony, metallic, and/or icy bodies orbiting the sun.

Margaret K. Wetterer, *Clyde Tombaugh and the Search for Planet X.* Minneapolis: Carolrhoda Books, 1996. Provides a detailed and compelling account of Pluto's discovery by a young novice astronomer from Kansas.

Index

Adams, John Couch, 6
ammonia, 19
asteroids
  described, 33
  Kuiper Belt and, 37–38
  as minor planets, 35
atmosphere
  composition of, 19–20
  of Earth, 15
  polar caps and, 21–22

blink comparator, 10–12

Ceres, 19
Charon
  discovery of, 24–26
  as KBO, 38
  orbit of, 26, 28
  origins of, 38–39
  rotation of, 28
  size of, 23–24, 27

Christy, James W., 24–26
composition, 19
core, 19

discovery
  of Charon, 24–26
  of Pluto, 10–13, 33–34
double planets, 23, 25–26, 27

Gill, Thomas, 9
gravity
  Neptune and, 6, 33–34, 36
  Triton and, 36
  Uranus and, 5–6, 33–34

Herschel, William, 5
Hubble Space Telescope (HST), 16

ice, 19

Kuiper Belt Objects
 (KBOs), 37–39

Leverrier, Urbain J.J., 6
Lowell, Percival, 9
Lowell Observatory, 9

Mercury, 19, 32
methane, 19, 21–22
naming
 of Charon, 26
 of Pluto, 13
Neptune
 gravity and, 6, 33–34, 36
 orbit of, 17, 33
 Pluto as satellite of, 35
nitrogen, 19

orbit
 of Charon, 26, 28
 definition of planets and, 32
 of Neptune, 17, 33
 of Pluto, 7–8, 14, 16–18, 21
 tilt of planets and, 17
 of Uranus, 5, 6, 33
origins
 of Charon, 38–39
 of Pluto, 35–37, 38

Pickering, William Henry, 8
planetesimals, 38
Planet O, 8
planets
 asteroids as, 35
 definition of, 31–32
 double, 23, 25–26, 27
 in solar system, 4
 tilt of, 17
Planet X, 9, 33–34
polar caps
 atmosphere and, 21–22
 composition of, 19

rotations, 28–30

size
 of Charon, 23–24, 27
 definition of planets and, 32–33
 of Pluto, 18–29, 32
solar system
 moons in, 32–33
 planets in, 4

temperatures, 14–15, 21
Todd, David Peck, 6–8
Tombaugh, Clyde W., 10, 11–13, 33
Triton, 33, 35, 36

Uranus, 5–6, 33–34

46   Pluto

Picture Credits

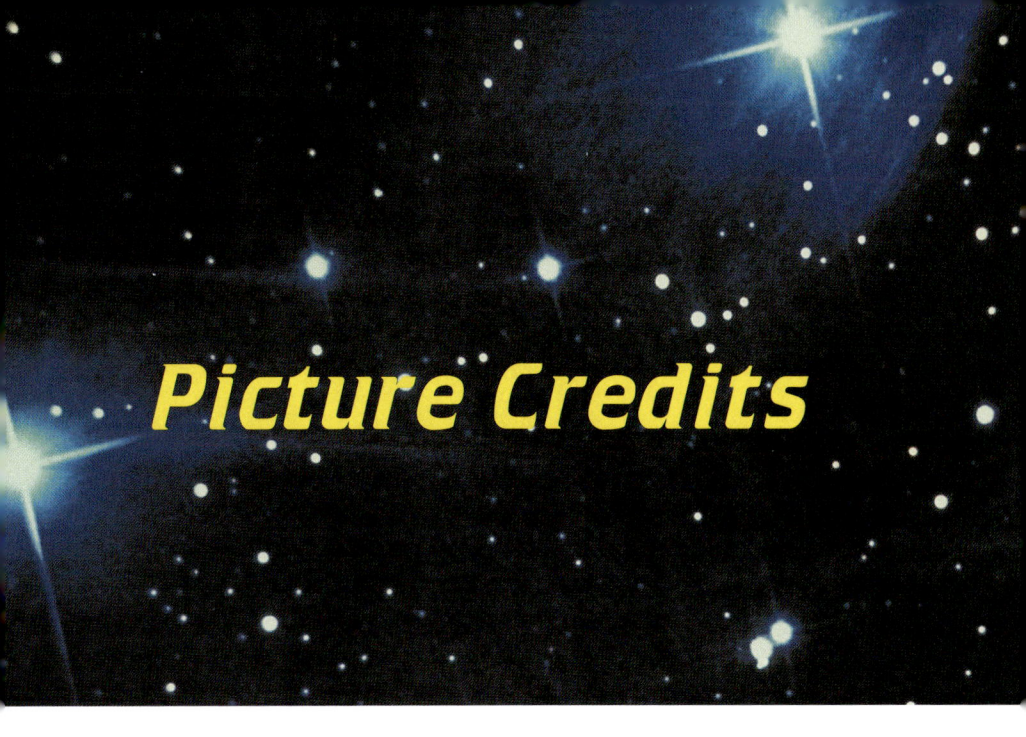

Cover and Title Page photo: © Lynette Cook/Photo Researchers

© S. Bensusen/Photo Researchers, 38

© Lynette Cook/Photo Researchers, 12

© CORBIS, 24, 25

Jeff DiMatteo, 18, 20, 28

© Mark Garlick/Science Source/Photo Researchers, 32, 34

© David A. Hardy/Science Source/Photo Researchers, 29

Chris Jouan, 37

Chris Jouan and Martha Schierholz, 5, 39

NASA, 15, 36

© NASA/Photo Researchers, 16

© Scala/Art Resource, 26

© Science Photo Library/Science Source/Photo Researchers, 8, 11

© Space Telescope Science Institute/NASA/Science Source/Photo Researchers, 7, 21

About the Author

In addition to his acclaimed volumes on ancient civilizations, historian Don Nardo has published several studies of modern scientific discoveries and phenomena. Among these are *The Extinction of the Dinosaurs, Cloning, Atoms, Germs, Neptune,* and *The Moon,* and a biography of the noted scientist Charles Darwin. Mr. Nardo lives with his wife, Christine, in Massachusetts.